What Is It Made Of?

Molly Finnegan

New York

What is it made of?
It is made of paper.

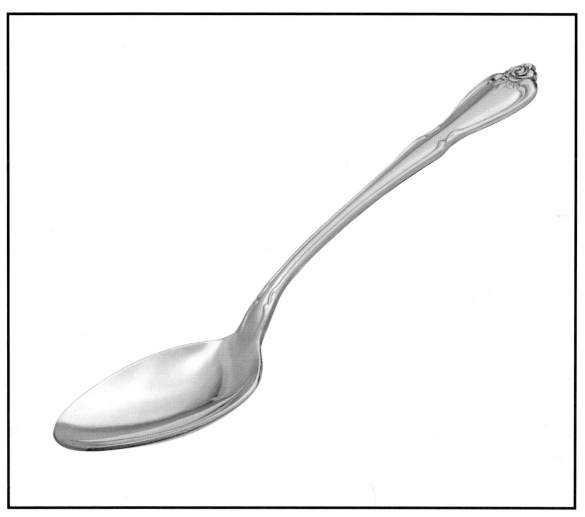

What is it made of?
It is made of metal.

What is it made of?
It is made of clay.

What is it made of?

It is made of cloth.

What is it made of?
This house is made
of wood!

object	is made of
	paper
	metal
	clay
	cloth
	wood

Words to Know

clay

cloth

metal

paper

wood